Units of Measurement

Sizing Things Up

Glen Phelan

Sally Ride, Ph.D., President and Chief Executive Officer; Tam O'Shaughnessy, Chief Operating Officer and Executive Vice President; Margaret King, Editor; Monnee Tong, Design and Picture Editor; Erin Hunter, Science Illustrator; Brenda Wilson, Editorial Consultant; Matt McArdle, Editorial Researcher

Program Developer, Kate Boehm Jerome
Program Design, Steve Curtis Design Inc.
www.SCDchicago.com

Sally Ride Science
9191 Towne Centre Drive
Suite L101
San Diego, CA 92122

ISBN: 978-1-933798-70-7

Printed in the United States of America
10 9 8 7 6 5 4 3 2 1
First Edition

Cover: How do the members of a giraffe family measure up? A chart marked in meters shows their heights.

Title page: It's weigh-in time at a nature reserve in China. Measuring weight is a way to check on a baby panda's growth.

Right: People in the Netherlands use gauges like this one to measure water levels. Flooding is a threat in their low-lying country.

Contents

Introduction

In Your World

The scientists in this sub are making an amazing journey—a voyage to the bottom of the sea! It's not the most comfortable trip. Space is tight. This French mini-sub, *Nautile,* is loaded with computers, cameras, and other equipment. There's barely room left for three people.

Nautile was designed to be small. That way a ship can carry it out to sea. Plus, a small sub can move easily among the mountains and canyons on the bottom of the ocean.

Size is important to the crew. So is time. There's only a certain amount of oxygen in *Nautile*'s tanks. A dive lasts 5 to 8 hours. *Nautile* may take 2 hours to reach the seafloor and 2 hours to surface. Scientists must make good use of their limited time to explore the ocean floor.

How small? How deep? How much time? *Nautile*'s crew depends on measurements. You do, too. Let's enter the world of rulers and clocks to see how.

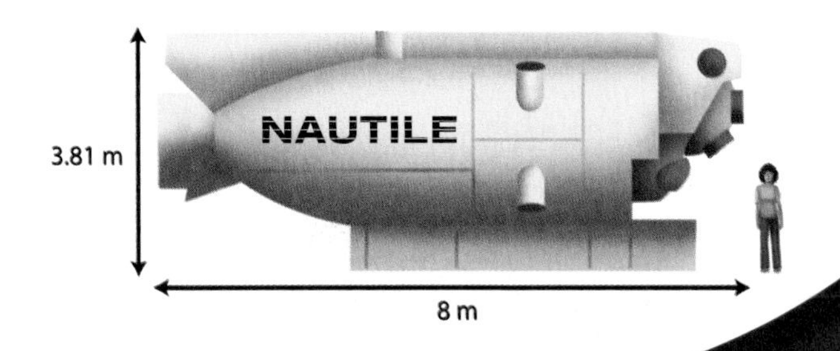

Choose Your Tool

How tall is a skyscraper? How heavy is a polar bear? How long will the movie in science class last? The answer to all three questions might be the same—1,000. Does that answer make sense to you? It shouldn't, because something very important is missing—the units! A unit gives meaning to a number. It answers the question *1,000 what?* So what units make sense in these examples? A skyscraper could be 1,000 *feet* tall. A polar bear might weigh 1,000 *pounds*. A movie in class could last 1,000 *seconds*, or about 17 minutes.

Now the number 1,000 makes sense. You know about how long a foot is, what a pound feels like, and how long a second lasts. These units don't change, and they are used by many people. That makes them standard units.

What's 1,000 feet tall? Try a skyscraper. At 1,053 feet, the Burj Al Arab hotel in Dubai, United Arab Emirates, is one of the tallest hotels in the world. The eye-catching design was inspired by a sailing ship.

Lost in Space

It's not good enough to use just *any* standard units in a measurement. You have to make sure you are using the *correct* units. That's a lesson even experts sometimes forget.

In September 1999, a spacecraft reached Mars after a 9-month journey. The *Mars Climate Orbiter* was going to study dust storms and weather on the Red Planet. But it never got the chance. Instead of going into orbit around Mars, the spacecraft burned up in the Martian atmosphere. What happened?

A mix-up in units of measurement doomed the *Mars Climate Orbiter.*

Engineers on Earth sent commands to the *Orbiter* to fire its thrusters, or small rocket engines, as it approached Mars. The thrusters fire with a certain amount of **force**. Here's the thing about force—you can measure it using units called pounds or units called newtons. The units are from two completely different systems of measurement. In fact, 1 pound of force is the same as about 4.5 newtons of force. The numbers the engineers sent to the *Orbiter* were in pounds. Guess what units the onboard computer was expecting? That's right—newtons.

Oops! The mixing of the different units sent the *Orbiter* spacecraft off course. The mission was over.

The Bottom Line

Any measurement includes a number and a unit.

A Flying Leap

The long jumper races down the track, leaps, and lands heels first in the sand. Did she jump far enough to win? There's only one way to find out. Measure it!

The jump was 23 feet 6 inches. It was also 7 meters 16 centimeters. How can the same distance have two different measurements? The first is in the **U.S. measurement system**—the one used in the United States but almost nowhere else. This system uses inches, feet, yards, and miles to measure distance and length. Here's how these units are related.

1 foot = 12 inches
1 yard = 3 feet
1 mile = 5,280 feet

If this were an international competition, 7 meters 16 centimeters would flash on the scoreboard. This measurement is in **SI** (*Système International d'Unités*, in French), also called the metric system. The SI units of distance and length include millimeters, centimeters, meters, and kilometers. These units are related, too.

1 centimeter = 10 millimeters
1 meter = 100 centimeters
1 kilometer = 1,000 meters

Could you jump a distance of 7.2 meters? You could probably make a good guess . . . if you know how long a meter is.

Why do some of the SI units have prefixes—those word parts in front of *meter*? The prefixes tell you what the unit means. *Centi-* means one-hundredth, or 1/100. So a centimeter is 1/100 of a meter, just like a cent is 1/100 of a dollar. *Kilo-* means a thousand. So a kilometer is 1,000 meters.

Dive Into SI

Thinking about lengths and distances in SI is a breeze, once you know how long the basic units are. A meter is a little longer than a yard. A centimeter, or 1/100 of a meter, is a little less than half an inch.

Let's dive into SI! Start with the width of this sea star. The photo shows the sea star's actual size—12 centimeters (4.7 inches) wide. Most of your classmates are about 1.5 meters (5 feet) tall. It may take you about 15 minutes to walk 1 kilometer (0.6 miles).

Now think about this—does it make sense to measure the distance between cities in centimeters? Would you measure the length of a bee in kilometers? If you answered "no" to both questions, then you're thinking in SI! The first step for any measurement is choosing the right unit. You use longer units, like kilometers, to measure longer lengths and distances. Shorter units, like meters or centimeters, are perfect for shorter lengths and distances.

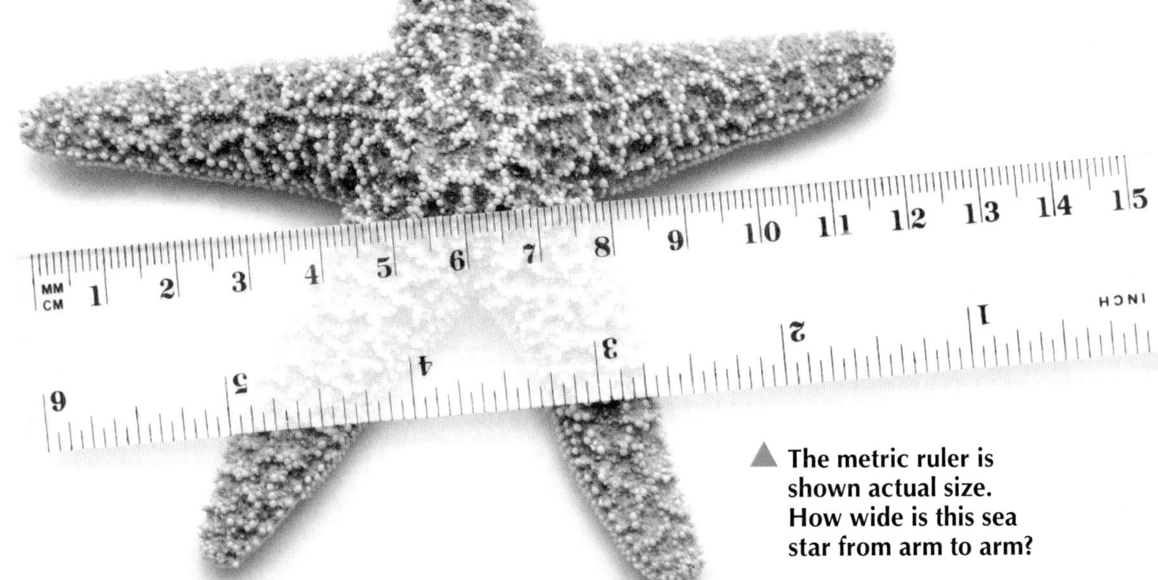

The metric ruler is shown actual size. How wide is this sea star from arm to arm?

The Bottom Line

Centimeters, meters, and kilometers are SI units used to measure length and distance.

A playground teeter-totter is kind of like a scientific measuring device called a balance. When two children are perfectly balanced, what does that tell you?

A Balancing Act

Imagine two kids on a teeter-totter. Instead of bouncing up and down, they sit still at either end of the board, their feet dangling in the air. The board stays level. What does this show about the two kids?

The balanced teeter-totter shows that the two children have the same **mass**. That means they both have the same amount of **matter**. Everything you see—from the dust bunnies under your bed to the planets and stars in space—is matter and has mass. Even some things that you can't see, like the air you breathe or the helium that fills a balloon, are matter and have mass.

How can you tell how much mass something has? The teeter-totter gives you a clue. You can measure the mass of small objects using a **balance**—a sort of scientific teeter-totter.

The object you want to measure goes on one side of the balance. Then you place objects called masses on the other side until the pans are even, or balanced. The masses are in SI units called grams. Larger, heavier objects are measured in kilograms. One kilogram is 1,000 grams. But you already knew that because of the prefix *kilo-*, right?

Weight a Minute!

With all this talk about mass, you may be wondering, don't you really mean **weight**? Don't kids balance on a teeter-totter because they *weigh* the same? Well, they do, but they weigh the same because they have the same mass.

Weight is a measure of the force of **gravity** pulling on an object. The more mass an object has, the more gravity pulls on it and the more it weighs.

A spring scale measures weight. You've probably seen spring scales in the fruits and vegetables section of a grocery store. Bathroom scales are spring scales, too. Can you figure out the SI unit for weight? Remember, weight is a measure of force. And like other forces, weight is measured in SI units called newtons. In the U.S., we use the units pounds and ounces.

A spring scale shows that this banana weighs 5.8 ounces, or a little over one-third of a pound. That's the same as 1.6 newtons in the metric system.

The Bottom Line

The SI units for measuring mass—the amount of matter an object has—are grams and kilograms.

At times you may wish your classroom clock could speed up or slow down. But you know that will never happen—the clock will just keep ticking at the same pace.

Ticktock, Ticktock

Have you ever watched time tick by on your classroom clock? Maybe you were counting down the seconds until after-school swim practice, hoping time would speed up. Maybe you were dreading an upcoming oral report, hoping time would stop. But no matter how you wished, the seconds ticked by at a steady, even pace.

Time is like that. You can't rush it. You can't stop it. But you sure can measure it. The unit for measuring time is the same in SI as it is in the U.S. system—the old reliable second.

In a way, a second is like a penny. It's not much by itself, but it adds up. Sixty seconds make a minute. Sixty minutes make an hour. Twenty-four hours make a full day. These bigger chunks of time come in handy. After all, would you rather say, "See you tomorrow" or "See you in 57,600 seconds"?

A Walk Through Time

Imagine a world without clocks or watches. How would you be able to tell time? Before clocks were invented, people measured time by observing the position of the Sun in the sky. They built sundials and other structures that used shadows to tell time. As the Sun's position in the sky changed throughout the day, so did the shadows.

Another early timekeeper was the water clock. It had a container shaped like a funnel with a tiny hole at the bottom. Water dripped through the hole into a bowl. Marks on the bowl showed hours as water filled to each level.

Over the years, people invented clocks that were more **accurate** and easier to use. In the 1600s, a scientist invented pendulum clocks, like grandfather clocks. Back then, clocks lost about a minute a day. Improvements made them accurate to within 10 seconds a day, then 1 second, then a hundredth of a second! As you'll see later, that's nothing compared to today's most accurate clocks.

▲ **Sundials can keep fairly accurate time . . . until nightfall. What else could keep them from working?**

The Sands of Time

Here's one ancient timing device you may have used while playing a board game—an hourglass! When all the sand flows through the opening, a certain amount of time has passed. Then you flip it over and start again. The flow might last an hour or a minute. It all depends on the size of the hourglass and the size of the opening.

The Bottom Line | **The second is the SI unit for measuring time.**

Is It Close Enough?

Suppose you are growing a plant in science class. It's time to see how tall your plant has grown. You hold the ruler against the stem and write down the height from the soil to the highest leaf—28.5 centimeters (11.2 inches). Your partner does the same thing—28.1 centimeters (11.1 inches). Why the difference?

Even though you both measured the same plant with the same ruler, you didn't measure it exactly the same way. Maybe your partner pushed the ruler into the soil a little bit. Maybe you didn't hold the ruler completely straight. Maybe you just weren't very careful when reading the smallest marks on the ruler.

That's okay—even pros make mistakes. In fact, carpenters have a rule—measure twice, cut once. It's better to double-check how long a piece of wood should be than to cut the wrong length. When else might double-checking a measurement come in handy?

▲ **Have you heard the saying "Measure twice—cut once"? It's a rule carpenters follow to make sure they measure accurately before they start cutting.**

Exactly Versus About

"You must be this tall to go on this ride." That's a familiar sign at amusement parks. You have to be a certain height to be safe on some thrill rides. This is a time when it's important for a measurement to be accurate.

When else does accuracy matter? How about clothing? If your shoes were 1 centimeter too short or too long, you'd probably notice. Even more exact measurements are needed for machines to work properly. A car won't get far if its hundreds of moving parts aren't exactly the right sizes. And a DVD has to be exactly the correct size or it won't fit in the DVD player, no matter how much you try to shove it in.

Sometimes, though, it's good enough just to **estimate**. You might spice up a pot of chili with a dash of hot sauce. There's no need to measure it exactly. You might tell a friend that you'll be at the park at about 1 o'clock. When else is *about* good enough?

Accuracy is important for a seamstress sewing a new blouse. If she doesn't measure carefully, one sleeve could turn out longer than the other.

The Bottom Line

We need accurate measurements for many things, but sometimes an estimate is good enough.

Right on Time

How much of your day depends on the time? Maybe more than you think. You know that if you aren't dressed and ready to go by a certain time, you'll be late for school. When are you meeting your friends to play basketball? Will you be home in time for your favorite TV show? How long will your homework take?

Time is important, all right, but most things in your life don't require accuracy to the second. If you get to the bus stop a minute earlier or later than usual, you probably will still catch the bus. Some things, however, depend on extremely accurate times. The last minute of a basketball game is measured in tenths of a second. The results of swimming and track meets are measured in hundredths of a second! A person holding a stopwatch can't measure time with this kind of accuracy, but computers can.

The difference between first and second place is sometimes a few hundredths of a second—faster than you can blink!

High-Tech Timing

Speaking of computers, they not only measure exact time, they depend on it. Computers are often connected in a **network** so people can share information. The biggest network of all is the Internet. You probably use it to find information, send messages, or listen to music. Well, the Internet wouldn't work if computers didn't know the exact time. You wouldn't receive a Web page or music file right after you requested it. And the times on Internet messages would be off. Try making sense of an email trail without having accurate times or knowing the correct order!

Another high-tech part of our lives that depends on accurate time is the **Global Positioning System** (GPS). People use this system of satellites to find exact locations on Earth. Each satellite sends its exact location and time to receivers on ships, planes, cars, and handheld devices. The exact time is important, because the receiver pinpoints where it is by analyzing the travel times of the satellite signals. The GPS is the best navigation tool since the compass!

Every Nanosecond Counts

The clocks on board each GPS satellite are accurate to within 1 billionth of a second, or a nanosecond! That kind of accuracy allows a person using GPS to pinpoint her or his location to within 30 centimeters (1 foot)!

▲ **Communication systems like cell phone networks depend on knowing the exact time to 1 millionth of a second, or a microsecond.**

The Bottom Line

Today's computers and other high-tech gadgets depend on extremely accurate time measurements.

Chapter 3: The Standard Units

It's Official

Lay an arm on your desk with your palm down. Now check out the distance from your elbow to the tip of your middle finger. This is called a cubit. The ancient Egyptians and other early cultures used the cubit as a unit of length. It wasn't a very good one, though. Can you tell why?

Everyone's body is different. So the length of a cubit is different from person to person. This caused confusion long ago when people were buying things like plots of land or pieces of lumber. People needed a set of standard units so that lengths could be measured the same way every time.

Ye Olde Measurements

▼ **In the past, measurements were based on things that could vary.**

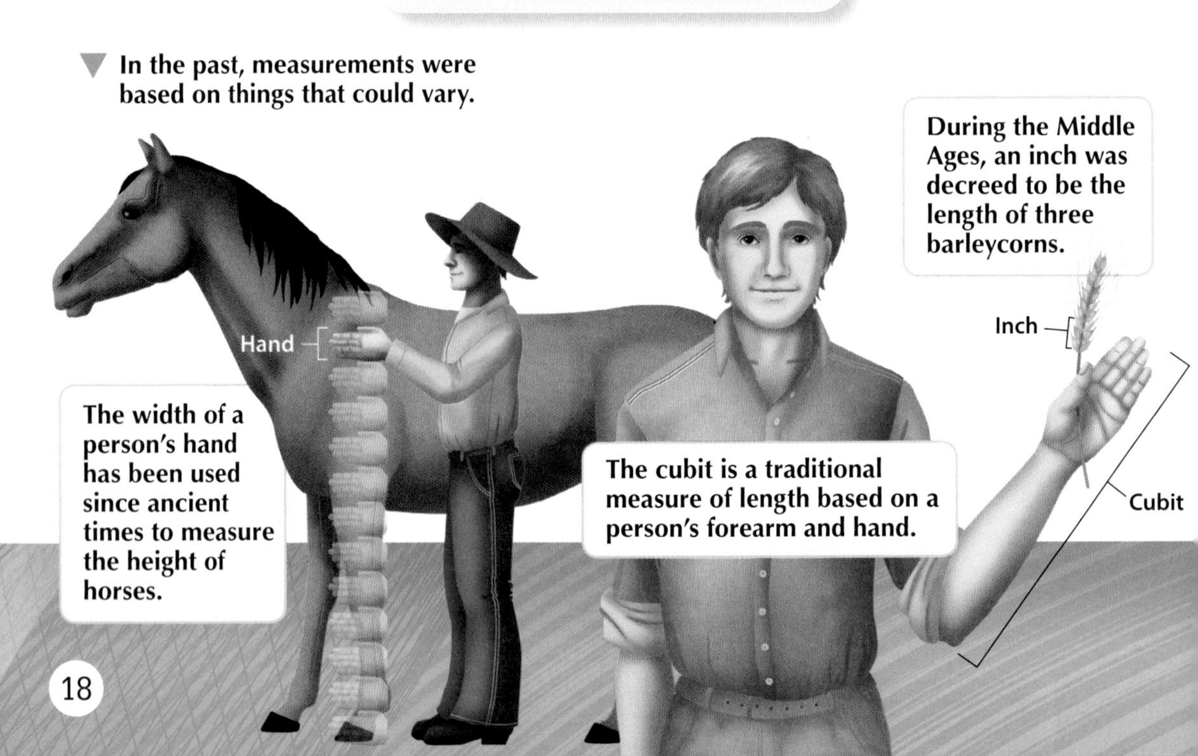

Setting a Standard

Scientists thought the unit of length should be based on something unchanging. You'd never guess what they came up with. In 1791, scientists meeting in Paris, France, decided the standard unit would be one ten-millionth of the distance from the North Pole to the equator. They called that fraction of distance the meter. A metal meter bar was made to serve as this official unit of length in the new metric system.

Today's technology has allowed scientists to define the meter more accurately. But the latest definition may sound just as bizarre as the first. The meter is the distance light travels in 1/299,792,458 of a second. Oh, and the light has to be in a **vacuum**—a tube without air—so the light isn't slowed by passing through air.

The length of a meter hasn't changed. The things that have changed are the official way the length is measured and how incredibly accurate the measurement is.

One of the first official meter bars is kept at the International Bureau of Weights and Measures near Paris, France.

North Pole
Dunkirk
Barcelona
Equator

The Wow!

A Dangerous Mission

How do you figure out one ten-millionth of the distance from the North Pole to the equator? First you measure part of the route. Then you do the math to calculate the entire distance and then one ten-millionth of it. In 1791, surveyors began measuring the distance from Dunkirk, France, to Barcelona, Spain. It was dangerous work. France and Spain were on the brink of war. The surveyors were arrested as spies and almost lost their heads!

The Bottom Line **Technology is used to define the meter as an exact length.**

Who Cares?

Who cares how a meter is measured? You do! You just may not realize it.

The high-tech definition of a meter is important in two big ways. First, it's extremely accurate. That allows for greater accuracy in maps and in pinpointing your location using GPS.

The accuracy of the meter was important in building the Channel Tunnel, or Chunnel, which connects England and France. The Chunnel is about 50 kilometers (31 miles) long. It goes through rock beneath the water of the English Channel. Workers started drilling from England and from France.

The fact that the meter is measured so accurately guaranteed that the two tunnels would meet up in the middle to form the Chunnel!

The new definition of the meter also is important because it means the meter can be recreated in any well-equipped lab or factory. So clothes can be made to fit better and the parts of do-it-yourself furniture can be cut so that they go together easily.

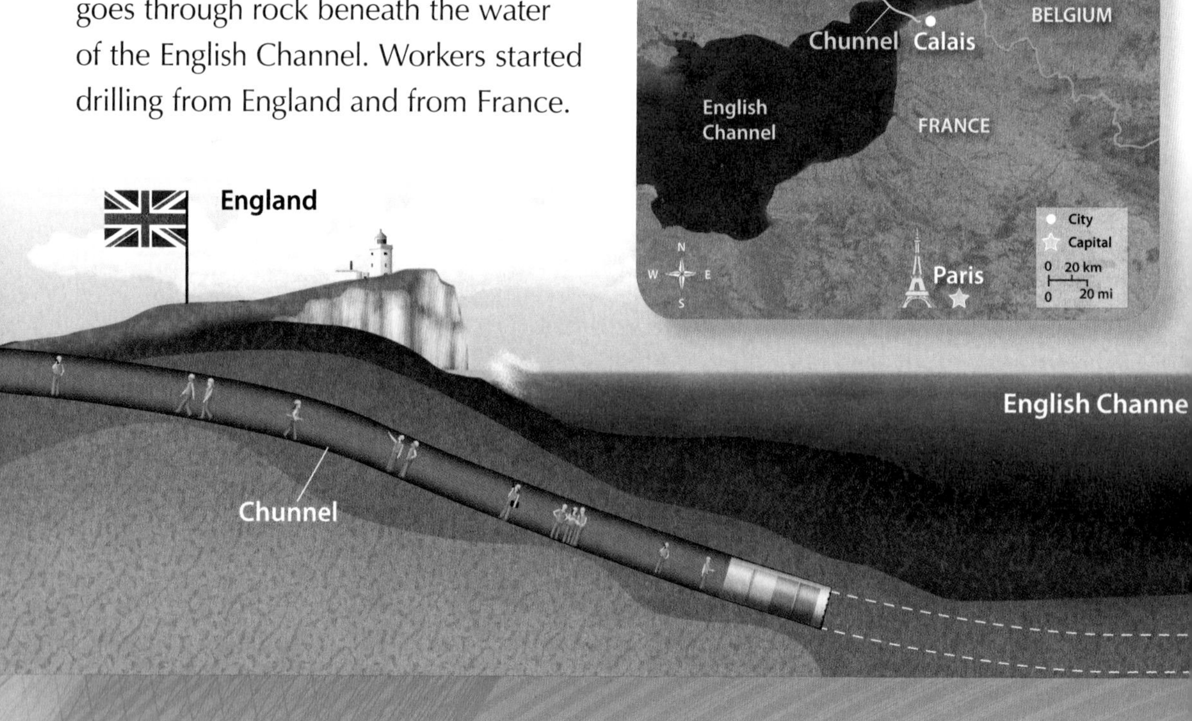

Under Glass

Imagine you're visiting the International Bureau of Weights and Measures, near Paris, France. Sparkling ponds and colorful gardens surround a beautiful building. But what you want to see lies in the basement. There's a vault down there. You need three keys to open it. A building engineer carefully controls the **temperature** and **humidity**. Sitting on a platform in the vault is a series of glass domes, each enclosing a smaller one. There beneath the smallest dome is the reason for all this special treatment.

▲ **Talk about special treatment—a series of glass domes protects the official kilogram mass.**

It's a cylinder of metal—but not just any cylinder of metal. It's the official mass of a kilogram, made in 1879. Other countries have copies of it that serve as their official standard unit for mass.

You can't get your hands on that cylinder, but if you want to know what a kilogram feels like, just lift a liter of water to your lips. That's about a kilogram.

France

▼ **Imagine drilling from both sides of the English Channel . . . and missing! Careful measurements and an accurate meter made sure that didn't happen.**

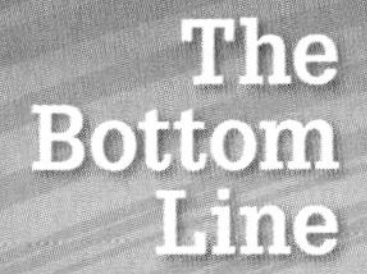

The kilogram is defined as the mass of a particular metal cylinder.

One Mississippi, Two Mississippi . . .

That's a popular way to count seconds out loud. As you might guess, though, the time it takes to say "Mississippi" is not the official definition of a second.

The second used to be defined as 1/86,400 of a day. That makes sense—60 seconds times 60 minutes times 24 hours equals 86,400 seconds. A complete day is when Earth spins once around its **axis**. Generally this takes 24 hours. But Earth wobbles a bit as it rotates, so a day might be a fraction of a second longer or shorter than 24 hours.

So what's the big deal? Because the length of a day changes, the length of a second would change, too—if a second were defined as a fraction of a day. Think of it this way. You and a friend are going to split the money you earn from selling your stuff at a garage sale. You each get half. On Thursday you earn $10, so you each get $5. On Friday, you earn $11, so you each get $5.50. On Saturday, you sell $16 worth of stuff, so you each get $8. The fraction that you get is the same, one-half, but the amount is different because the *total amount* is different.

▲ The Sun rises and sets as Earth spins on its axis. A day is the time it takes for our planet to rotate once. But because Earth wobbles a little as it spins, the length of each day varies slightly.

In the same way, the length of 1/86,400 of a day changes as the total length of the day changes. A standard unit of measurement that keeps changing is not a standard at all! A better way of defining a second was needed.

In the 1950s, scientists found a better way. They invented **atomic clocks**. These devices measure how quickly an electron in a cesium atom vibrates. How many times does the electron vibrate every second? It's another long number—9,192,631,770 times!

▲ **This atomic clock at the National Physical Laboratory in England is accurate to within a second every one million years.**

A Timely Contest

The need for accurate time didn't begin with the computer age. In the 1700s, measuring time precisely could be a matter of life and death. Ships often smashed into rocks at night because navigators didn't know their exact longitude—their location in an east-west direction. The British government offered a reward for a device that could find longitude at sea.

An English carpenter named John Harrison invented a clock that kept accurate time even on rough seas. A navigator could use the clock to figure out how many degrees of a circle a ship had traveled. The clock was off by only a couple of seconds on a monthlong voyage, so the ship would be off course by less than a mile. That was good enough to make ocean voyages safer but nowhere near as precise as today's super-accurate timepieces. What would 18th century sailors think of atomic clocks and GPS?

The Bottom Line

Atomic clocks keep accurate time by measuring the vibrations of electrons.

THINKING LIKE A SCIENTIST

When is an error not a mistake? When it's an error in measurement! Scientists make measurement errors all the time. In fact, every measurement has some error.

Measuring

No measurement is exact. Every measurement is made to the nearest something. The smaller the unit on the measuring tool, the more accurate the measurement can be. For instance, you could measure the length of your desk using a ruler that is marked in centimeters and millimeters. So you could measure to the nearest millimeter. But maybe the edge of the desk is somewhere between two millimeter marks. Even with great eyesight, you'd be estimating the true length.

You may be getting the idea that measuring isn't as simple as it seems. If you measure something three times, you might get three different measurements. How can you make the most accurate measurement possible? Find an average.

You find an average of a set of numbers by adding them together to get a total and then dividing the total by how many numbers are in the set. A scientist measured the length of the snow leopard track in the photo three times. The measurements were 10.2 centimeters (cm), 10.3 cm, and 10.4 cm. Find the average.

$$10.2 \text{ cm} + 10.3 \text{ cm} + 10.4 \text{ cm} = 30.9 \text{ cm}$$

$$\frac{30.9 \text{ cm}}{3} = 10.3 \text{ cm}$$

The scientist recorded a length of 10.3 centimeters in her notebook.

▶ **How can you make sure a measurement of something—like this snow leopard track—is as accurate as possible?**

Interpreting Data

Scientists use data tables to organize and analyze the information they collect. Look at this table. The data show the length of the stride of a black bear when it is walking. This is the distance from the tip of a toe of one front foot to the tip of the toe of the back foot on the same side. The scientist measured two sets of bear tracks. She measured the stride seven times for each set. Here is her data.

Stride of Black Bear Tracks (cm)

Set 1	Set 2
45.6	27.6
45.5	27.5
45.7	27.8
45.7	27.6
45.9	27.9
45.8	27.9
45.8	27.8

Your turn! Use the information on this page to answer the following questions.

1. What is the shortest stride in Set 1?

2. What is the longest stride in Set 2?

3. Find the average stride for each set. Round the average to the nearest tenth.

4. Why do you think the two sets of tracks show strides of two very different lengths?

THE ISSUE

The Measure of a Forest

Forests are full of trees—and carbon. Trees absorb lots of the element carbon as carbon dioxide, or CO_2, from the air. Trees use that CO_2 during photosynthesis to make sugar for food. They also use the carbon to make fats and proteins. These carbon-carrying molecules make up the cells of tree trunks, branches, and leaves. In fact, carbon makes up about half the dry weight of a tree. When trees burn or decay, they release that carbon—as CO_2—back into the air. But how much carbon?

One way to measure carbon is . . .

. . . by chopping down, drying, and weighing all of the trees in a forest. But that's way too destructive. Instead, if you accurately measure the heights of the trees, you can estimate the diameter and mass of each one. And if you know their masses, you can estimate the dry weight and carbon content of all those trees! Over time, the measurement shows if a forest is growing or shrinking. It also reveals whether the forest is soaking up CO_2 from the atmosphere—or adding CO_2 back to it.

▲ Trees—like these mangroves—absorb CO_2, a greenhouse gas that traps heat and increases global warming.

Lola Fatoyinbo-Agueh

"We've lost 30 percent of the world's mangroves in the last 20 years alone," says Lola. "That's more, as a percentage, than the losses of rainforest or coral reef."

From the time she was young, Lola Fatoyinbo-Agueh was fascinated by mangrove trees. She spotted mangroves on family trips along the coast of West Africa. The trees' crooked roots grow up and out of salty lagoons. "It's like they are on stilts—half in the water, half out, like a floating forest," says Lola.

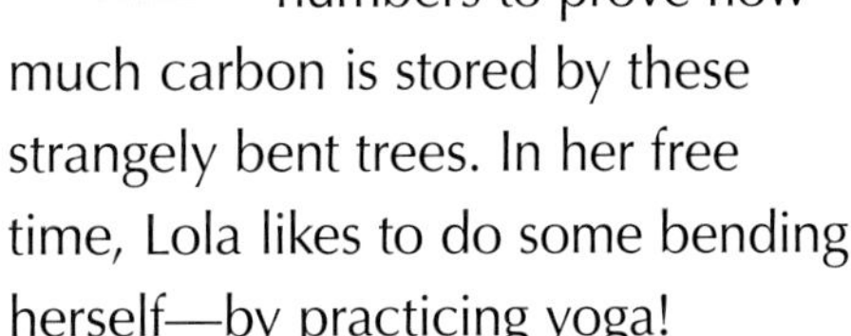

Lola saw how pollution and construction were destroying the mangroves. "It really broke my heart," she says. That inspired her to study ecology—with a focus on saving the world's mangroves—when she came to the United States.

Mangroves are an important habitat for hundreds of species of fish, birds, snakes, and other organisms. Lola's work also shows how important mangrove forests are for storing carbon. Mangroves store as much carbon as rainforests—about 45 kilograms (100 pounds) an acre per day! In her job, Lola gathers the numbers to prove how much carbon is stored by these strangely bent trees. In her free time, Lola likes to do some bending herself—by practicing yoga!

IN THE FIELD

▲ **A student measures the distance to a mangrove trunk before calculating its height with a clinometer like the one below.**

Squelch! Lola Fatoyinbo-Agueh and her students slog across the small and muddy island off the coast of Mozambique, in southeastern Africa. Lola pauses. "Time to measure another mangrove," she says.

"This one will be number 1,978," a student notes, carefully measuring Lola's distance to the tree. Lola imagines a horizontal line between her and the trunk at eye level. She then uses a device called a clinometer to measure the angles between that line and two other imaginary lines—from her eye to the top of the tree and from her eye to the base of the tree.

Knowing these measurements allows Lola to calculate the tree's height. "It's 4.6 meters," she says. A student records the height. "With these hand measurements, we'll confirm the accuracy of satellite measurements made from space," Lola explains.

"And if they check out, we can quickly size up the mangrove forests across all of Mozambique—and Africa!" a student adds.

TECHNOLOGY

Lola and other scientists measure tree heights using a satellite 600 kilometers (370 miles) above Earth. Now that's *high*-tech! An instrument aboard the satellite bounces laser pulses off the Earth's surface below. To find out how far the laser pulses traveled, scientists keep track of how long each bounce takes. When a pulse hits an object such as a tree, the pulse takes a little less time to bounce back to the satellite. That difference shows how tall the tree is. It's almost like stretching a measuring tape from the satellite to Earth, over and over.

Investigation Connection

Measure Up

You can try this low-tech way to measure a tree's height. Fold a square piece of paper in half to form a triangle. Hold the triangle up to your eye like in this illustration. Carefully take steps backward until the top of the tree lines up with the top of your triangle. Mark your location. Then carefully measure the distance to the base of the tree in meters or feet. That distance—plus your height—gives you a rough idea of the tree's height.

A

B

A + B

Oh, By the Way

African coastal areas are dense with mangroves—shown in green in this satellite image. Mangroves protect against storms and erosion, and take in carbon. Mozambique alone has about 2,900 square kilometers (1,120 square miles) of mangroves. That's more than three times the area of New York City!

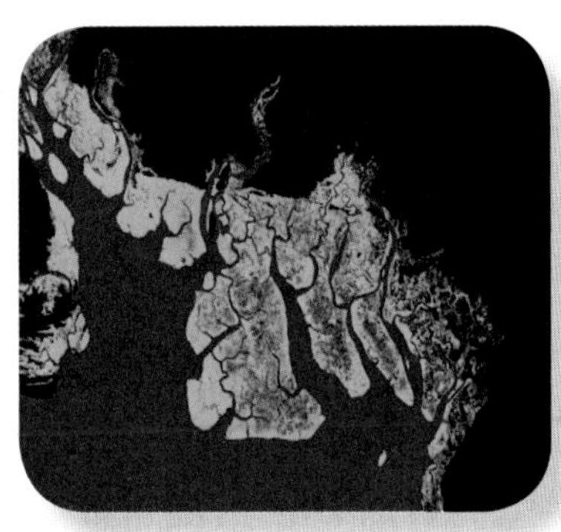

Hey, I Know That!

You've learned a lot about units of measurement, from ancient cubits to modern meters. You've learned what the units are and what they mean. And you've learned that sometimes close is good enough, but at other times, a measurement must be precise.

On a sheet of paper, show what you know as you do the activities and answer these questions.

1. What is the length of the toy truck shown here in millimeters? In centimeters? In meters? In kilometers? (pages 8 and 9)

2. You are measuring the length of a room. Would you use a meter stick or a metric ruler? Explain your choice. (page 9)

3. Which of these measurements make sense? For each one that doesn't make sense, change the unit or the number so that it does. (pages 8-12)

   a. a bird's wingspan—40 meters
   b. the mass of a pumpkin—8 kilograms
   c. the distance between your home and school—1 kilogram
   d. length of lunchtime—1,800 seconds
   e. length of your favorite TV show—10,500 seconds
   f. height of this book—23 centimeters

4. Choose a topic in the book. It might be the difference between the U.S. and SI systems, the history of timekeeping, or something else. Make a science poster that explains the topic to others. Include pictures, graphs, and captions.

Glossary

accurate (**adj.**) free from error or agreeing closely with a standard value (p. 13)

atomic clock (**n.**) an extremely accurate clock whose time is controlled by steady processes in atoms or molecules, such as vibrations. The standard atomic clock is based on the vibrations of cesium atoms and is so accurate that it would gain or lose less than one second in one million years. (p. 23)

axis (**n.**) the imaginary line about which an object, such as a planet, rotates (p. 22)

balance (**n.**) a tool used for measuring an object's weight—that is, its mass under the influence of gravity (p. 10)

estimate (**v.**) to approximate the size or other characteristics of something (p. 15)

force (**n.**) a push or pull applied to an object (p. 7)

Global Positioning System (**n.**) a navigation system, abbreviated GPS, that uses dozens of satellites in space to allow people anywhere in the world to determine their exact location and time of day. The satellites are spaced so that from any point on Earth, four satellites are above the horizon. Each satellite contains a computer, an atomic clock, and a radio, and each continually transmits its changing position and time. Anyone with a GPS receiver can find out her or his precise location, or longitude and latitude. (p. 17)

gravity (**n.**) the attractive force that any object with mass has on all other objects with mass. The greater the mass of the object, the stronger its gravitational pull. (p. 11)

humidity (**n.**) a measure of the amount of moisture or water vapor in the air (p. 21)

mass (**n.**) a measure of the total amount of matter contained within an object (p. 10)

matter (**n.**) any substance that has mass and takes up space (p. 10)

network (**n.**) a system of computers, computer accessories, and databases connected by communications lines (p. 17)

SI (**n.**) (from French: *Système International d'Unités*) the metric system of units of measurement used by people around the world so that there is only one unit for each type of measurement. Fundamental units of measurement include length (meter), mass (kilogram), time (second), and temperature (kelvin). (p. 8)

temperature (**n.**) a measure of the average kinetic energy of the molecules in matter (p. 21)

U.S. measurement system (**n.**) a measurement system used less and less in the United States since the international adoption of the metric system in 1960. U.S. units of measurement include length (yard), mass (pound), and volume (gallon). (p. 8)

vacuum (**n.**) a region that contains no matter of any kind. Outer space is an example of a vacuum. (p. 19)

weight (**n.**) a measure of the force of gravity on an object, or weight = mass x acceleration of gravity (p. 11)

Index

About the Author Glen Phelan's fascination with science was sparked when he was a teenager by the lunar missions of the Apollo Program. He shares his fascination through teaching and writing. Learn more at www.sallyridescience.com.

Photo Credits carsthets/Shutterstock.com: Cover. Filip Fuxa/Shutterstock.com: Back cover. Katherine Feng/Minden Pictures: Title page. Ruud de Man/iStockphoto.com: p. 2. Alexis Rosenfeld/Science Photo Library: p. 4. hainaultphoto/Shutterstock.com: p. 6. NASA/JPL: p. 7. technotr/iStockphoto.com: p. 8. Joao Virissimo/Shutterstock.com: p. 9 (sea star). Stephen Aaron Rees/Shutterstock.com: p. 9, p. 30 (ruler). Fulvia Finelli/iStockphoto.com: p. 11. manley099/iStockphoto.com: p. 12. Alex Hubenov/Shutterstock.com: p. 13 top. Stuart Miles/Shutterstock.com: p. 13 bottom. Steve Cole/iStockphoto: p. 14. Diego Cervo/Shutterstock.com: p. 15. Reuters/Carlos Barria: p. 16. GPS.gov: p. 17 top. Cheryl E. Davis/Shutterstock.com: p. 17 bottom left. Supri Suharjoto/Shutterstock.com: p. 17 bottom right. Courtesy of the *Bureau International des Poids et Mesures*: p. 19 (meter bar), p. 21 (kilogram). Tom Patterson: p. 19 (globe in illustration). Galyna Andrushko/Shutterstock.com: p. 22. Andrew Brookes, National Physical Laboratory/Science Photo Library: p. 23. Eric Dragesco/Minden Pictures: p. 24. Eric Baccega/NPL/Minden Pictures: p. 25. Gusev Mikhail Evgenievich/Shutterstock.com: p. 26. Courtesy of Lola Fatoyinbo-Agueh: p. 27, p. 28. NASA: p. 27 (logo). Suunto: p. 28 (clinometer). NASA/Lola Fatoyinbo-Agueh: p. 29. Nikolai Tsvetkov/Shutterstock.com: p. 30 (truck).